CONTENTS

I. Chemistry of Benzoxazoles   1

II. Biological Activities of Benzoxazoles   31

.1. Benoxazoles As Anticancer Agents

2. Benzoxazoles As Antimicrobial Agents

3. Benzoxazoles As Anti-Histaminic Agents

4. Benzoxazoles As Anthelmintic Agents

5. Benzoxazoles As Hypoglycemic Agents

6. Benzoxazoles As Chymase Inhibitors

7. Benzoxazoles As 5-Ht$_3$ Agonists

8. Benzoxazoles As Hiv-1-Rt-Inhibitors

9. Benzoxazoles As Herbicidal Agents

10. Benzoxazoles As Potential Agonists Of Lh Hormones

11. Benzoxazoles As Uricosuric And Diuretics

12. Benzoxazoles As Steroid Sulfatase Inhibitors

13. Benzoxazoles As Inhibitors Of Immuno Complex
    Induced Inflammation.

14. Benzoxazoles Having Miscellaneous Activities

References   41

I. Chemistry of Benzoxazoles [1]

Benzaxozole (1) (m.p 27-30°; b.p. 182°C), is a planar molecule with conjugated π electrons sextets in the cyclic system. The chemical properties are aromatic in character. The lone pair of electrons on nitrogen, which is coplanar.

(1)

With the heterocyclic ring and therefore not involved in delocalization, confers weakly basic properties. Associated with the aromatic is a degree of stability, but when these are quarternized the resulting azolium species are significantly activated towards nucleophillic attack.

Benzoxazoles tends to react mainly at C-6 in electrophilic substitutions and to lesser extent at C-5. Nitration of benzoxazole affords the 6-nitro products (2).

(2)

Benzoxazoles are stable towards a range of reductive conditions, but reduction of the ring to oxazolidines can be effected with sodium in ethanol.

2-Arylbenzoxazoles undergo photo-Fries rearrangements. 2-Hydroxy Benzoxazoles (3) exist predominately in the 2-keto form (4).

(3)   (4)

Halogenobenzoxazoles (5) undergo a range of nucleophillic displacements which are summarized in the scheme.

Benzoxazole quarternizes to give the methiodide (6), but under more vigorous conditions may suffer ring cleavage.

(6)

Reaction of piperazines with 2-chlorobenzoxazole (7) result in the formation of compounds (8) and (9) [2]. Alkylation (9) with methyliodide results in the quaternary salt (10).

Similarly, compound (9) on allylation with allyl iodide in dimethyl formamide results in the corresponding quaternary salt (11).

2-Mercaptobenzoxazoles (12) exist predominantly in the thione form (13). Alkylation can occur at sulphur to yield compound (14).

Reaction of Z-X-Z (ex: X=CH$_2$, Z=Cl), with 2-mercaptobenzoxazole (12)

2 [benzoxazole]—SH + Cl-CH$_2$-Cl ⟶ [benzoxazole]—S-CH$_2$-S—[benzoxazole]

(12)   (15)

affords the corresponding thioether linked bis heterocycles[3](15).

Benzoxazoles are resistant to alkaline hydrolysis, but are readily cleaved by acids, probably because of nucleophillic attack.

[benzoxazole ⇌ benzoxazolium (N-H) ⇌ 2-(OH)-phenyl-NH-CHO]

Benzoxazole hydrolysis is relatively easy, 2-methyl benzoxazole (16) giving *o*-acetamido phenol in hot water, although the reaction is more rapid in dilute acid.

[2-methyl benzoxazole]—CH$_3$ $\xrightarrow{\text{H}_2\text{O/H}^+}$ 2-(OH)-phenyl-NH-CO-CH$_3$

(16)

Quaternary salts are hydrolysed more readily as shown by the hydrolysis of N-methyl benzoxazole (17).

[N-Me benzoxazolium] $\xrightarrow{\text{H}_2\text{O/H}^+}$ 2-(OH)-phenyl-N(CH$_3$)—CHO

(17)   (18)

5

Although 2-methylbenzoxazole (16) can be cleaved by methoxide at 120°C to give *o*-amino phenol (19).

$$\text{(16)} \xrightarrow{\text{CH}_3\text{O}^-/120^\circ\text{C}} \text{(17)}$$

(16)   (17)

Quaternization of benzoxazole (20) with methyl iodide at 120°C leads to 2-hydroxytrimethyl aniline.

$$\text{(1)} \xrightarrow{\text{CH}_3\text{I}/120^\circ\text{C}} \text{(20)}$$

(1)   (20)

The 2-amino benzoxazole (21) exists as the amine tautomers (22). The aminobenzoxazole (21) protonate on the ring nitrogen and reacts with methyl iodide at 100°C to give the N³-alkylated product (23). Reaction of 2-aminobenzoxazole with aroyl isothiocyanates gives N-aroyl-N′-(benzoxazolo-2yl) thioureas[4].The product (24) obtained on reaction with PCl₅ in POCl₃ and with oxidizing agents have been identified as 3-aroyliminobenzoxazolo [3, 2-b] [1, 2, 4]-thiazolidines(25).

(23)　　　　　(21)　　　　　(22)

(24)

(25)

Benzoxazole affords an excellent synthetic route to ω-nitrilo acids (28).

(26)　　　　　　　　　　　　　　　　　　　(27)

(28)

The reaction of *o*-aminobenzoxazoles with benzonitriles in the presence of anhydrous tin (IV) chloride gives N-(benzoxazol-2-yl)benzamidines in high yields. Cyclodehydrogenation of these compounds with lead (29) acetate affords 2-aryl[1,2,4]triazolo[5,1-b]Benzoxazoles (30) in good yields[5].

X=H, CH$_3$Cl
R=H, 2-CH$_3$, 4-CH$_3$, 2-Cl, 4-Cl, 4CH$_3$O-

Pb(OAc)$_4$/benzene
reflux,3hr
60-72%

(30)

6-chloro-2-phenoxybenzoxazole (31) reacts smoothly with a series of aliphatic primary amines to give 6-chloro-2-(substituted amino) Benzoxazoles (32), at room temperature, in presence of excess of amines reported by Jozsef Kover *et al*,[6] Thus, 2-phenoxy group of the benzoxazole system nucleophilically more susceptible that its 6-chloro group.

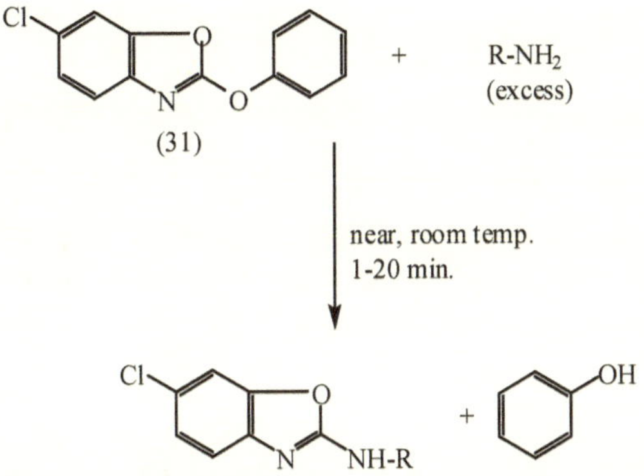
Cl
O
N
O
+
R-NH2
(excess)
(31)
near, room temp.
1-20 min.
Cl
O
N
NH-R
+
OH

(32)

2-Methylbenzoxazole (16) reacts with benzaldehyde in presence of Zinc chloride to give the 2-benzylidine derivative (33) indicating the reactivity of the methyl gourp linked to azo-methine system.

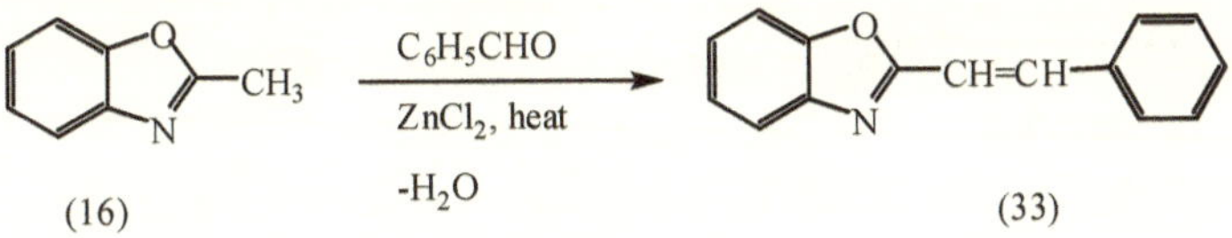
O
N
CH3
C6H5CHO
ZnCl2, heat
-H2O
O
N
CH=CH

(16) (33)

2-Benzylbenzoxazole (34) is reactive enough to couple with diazonium salts, and reacts at the 2-methylene group with aldehydes, nitroso compounds, and amyl nitrite.

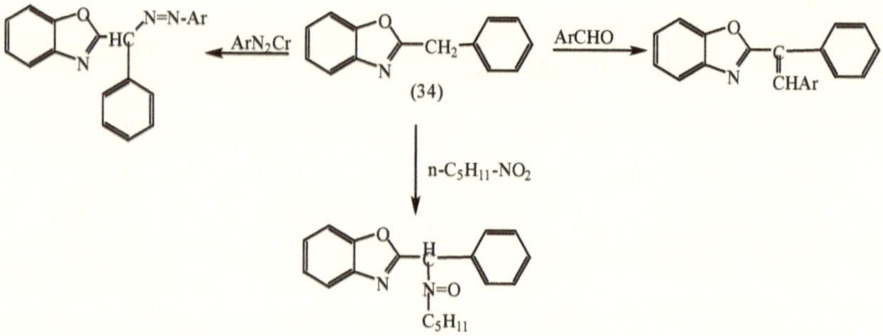

(34)

Benzoxazoles with acyl substituents at C-2 may undergo Grignard reactions (35).

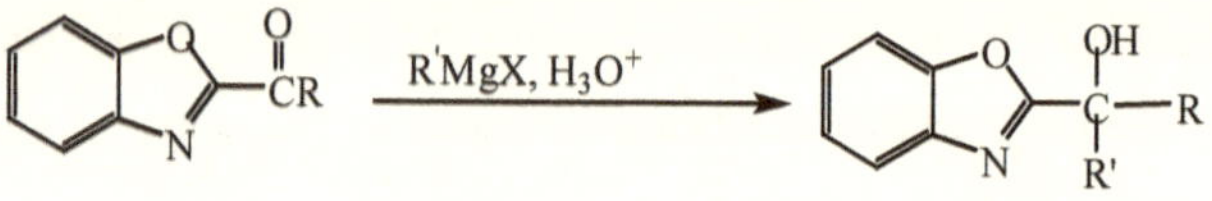

(35)

Synthesis of Benzoxazoles

The condensation of carbon-di-sulfide or cyanogen bromide with *o*-aminophenol leads to benzoxazole thione (36) or 2-aminobenzoxazole (37) respectively.

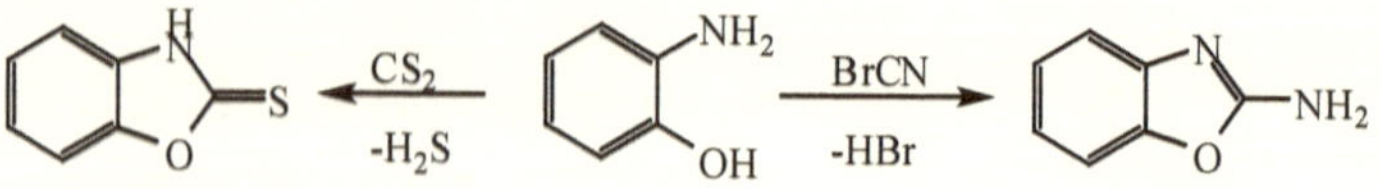

(36)   (37)

The Beckmann rearrangement of oximes *o*-hydroxybenzophenones leads directly to benzoxazoles (38).

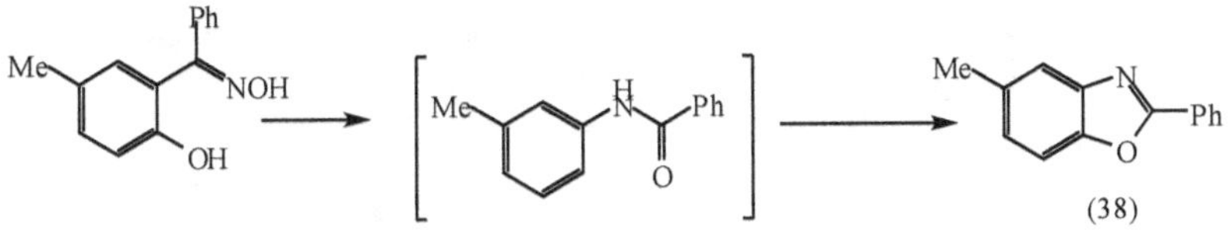

(38)

Benzoxazoles (39) and other condensed oxazoles are obtained by the oxidative ring closure of phenolic Schiff's bases.

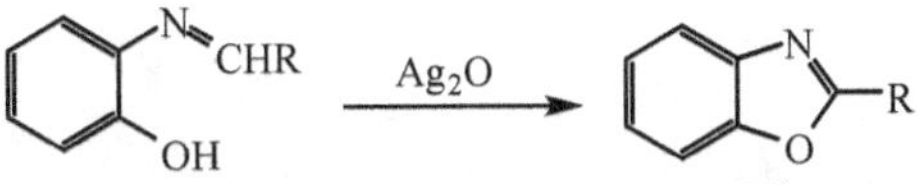

(39)

The formation of phenanthro oxazoles (40) by the action of benzylamine or other amines on phenanthraquinone.

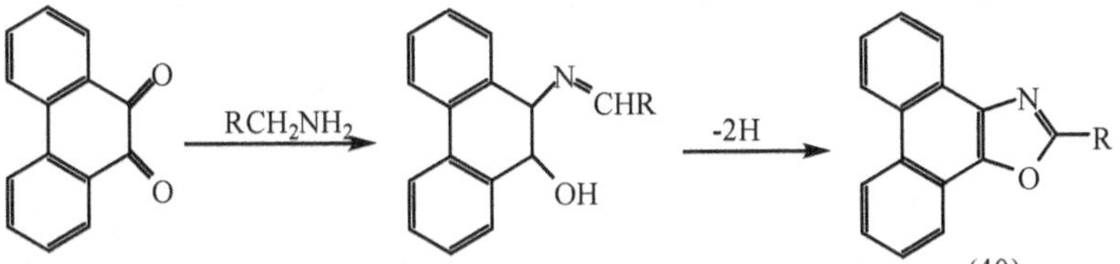

(40)

Thermal dehydration [7] of o-(acylamino) phenols is the method of choice for the preparation of Benzoxazoles (39). O, N-Diacyl derivatives of o-aminophenols cyclize at lower temperature than do the monoacyl compounds.

The synthesis is often carried out by heating the aminophenol with the carboxylic acid or derivative, such as acid chloride, anhydride, an ester, amide or nitrile.

11

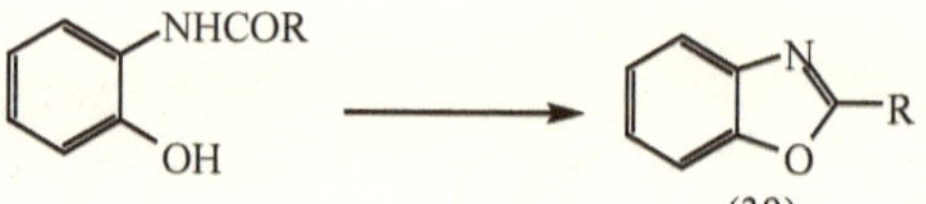

(39)

Thermal cyclization with acid catalysts is commonly employed to synthesize Benzoxazoles (1). For example, 2-amidophenol have been treated with PPA, PPE, Propionic acid, $POCl_3$ and $SOCl_2$ at high temperature to give Benzoxazoles.

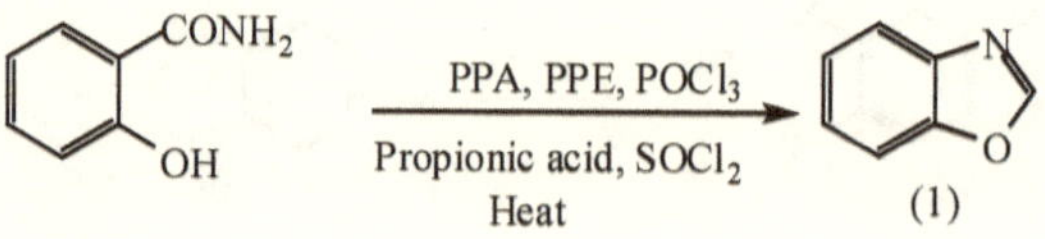

PPA, PPE, $POCl_3$

Propionic acid, $SOCl_2$

Heat

(1)

Benzoxazoles (40) have been obtained by heating o-aminophenol and carboxylic acids in the presence of PPA [8].

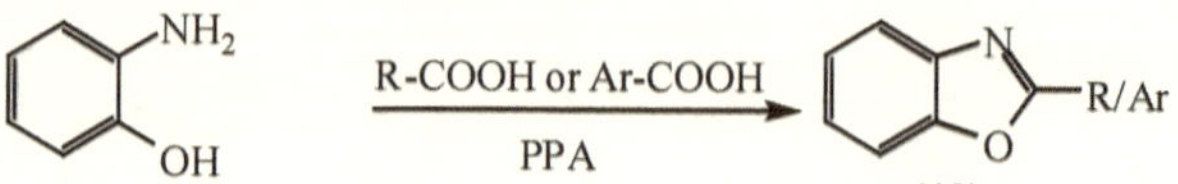

R-COOH or Ar-COOH

PPA

(40)

Benzoxazoles (41) are formed by the action of potassium amide in liquid ammonia on N-aroyl derivatives of both o- and m- chloroaniline. The reaction does not proceed directly from an intermediate aryne, as was once thought, but via isolable o-hydroxyphenyl amidines.

The 5 and 7-methylbenzoxazoles (42) are synthesized [9] by distilling appropriate aminocresol hydrochlorides with sodium formate.

R1 = CH$_3$, R$_2$= H
R$_1$ = H,   R$_2$ = CH$_3$

Various substituted 1,2-benzoxazoles (43) are prepared in good yields from salicylaldoximes and orthhydroxyphenyl ketoximes via intramolecular Mitsunobu reaction [10].

Bawal B.M et al. [11] reported the mild and simple method for the synthesis of Benzoxazoles (44) via Beckmann rearrangement of o-acylphenol oximes using zeolite catalyst.

$R^1 = CH_3$ or C_2H_5 $R^2 = CH_3$, H, Cl $R^3 = H$, OH

The synthesis of Benzoxazoles (1) by the cyclocondensation reaction of o-aminophenol with S-methyl isothioamide-hydroiodides on silica gel under microwave irradiation and also in solvent under reflux reported by Rostamizadeh et al. [12].

An advanced method involves cyclization of a 2-aminophenol with S-triazines reported by Grundmann et al. [13] triethyl ortho format [14,15] or isonitriles[16,17] and 2-hydroxybenzonitrile[18], photochemically reported by Ferris et al. to yield benzoxazole(1).

```
        |
        |        HC(OEt)₃
        |_____
        |        Isonitriles
        |_____
        |    2-hydroxy benzonitrile
        |_____
```

HC(OEt)$_3$

Isonitriles

2-hydroxy benzonitrile

2-(4-aryl-2-thiazolylamino) Benzoxazoles were synthesized in good yields [19]. Reaction of 2-amino-4-arylthiazoles with CS$_2$ and methyl iodide in DMF in presence of strong sodium hydroxide solution gives the corresponding dimethyl N-(4-aryl-2-thiazolyl) dithiocarbonimido dithioates. These compounds on reaction with 2-aminophenol in refluxing DMF, in presence of one equivalent sodium hydroxide affords the 2-(4-aryl-2-thiazolylamino)-benzoxazoles (45).

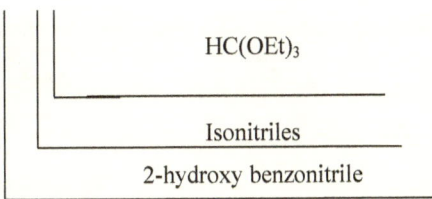

(45)

Piperidine-4-carboxylic acid and o-aminophenol were heated with polyphosphoric acid to afford [20] (46).

15

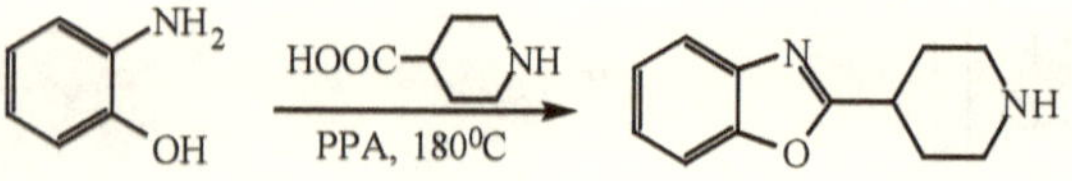

(46)

Naphtho-and Benzoxazoles (47) were synthesized and the method consist of a decomposition of naphtha and benzoxazinones with KOH reported by Claudiosatiz *et al.*[21]

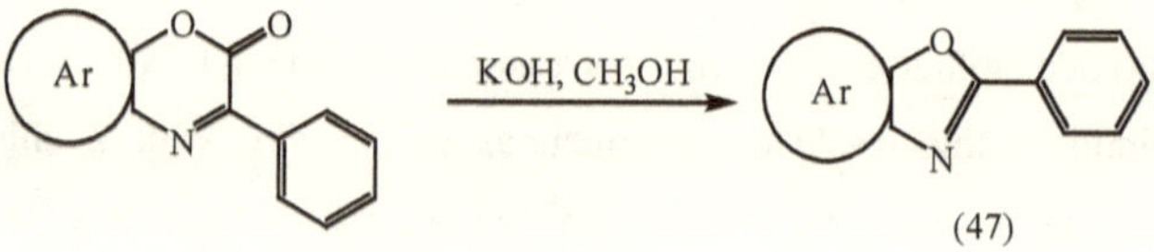

(47)

Ar = 1,2-benzo-, 1,2 - naphtho-, 2,1- naphtho-, 2,3-naphtho-

Synthesis of aromatic Benzoxazoles, which containing allyl ether groups (48, 49) reported by Thuy D.Dang *et al.*[22]

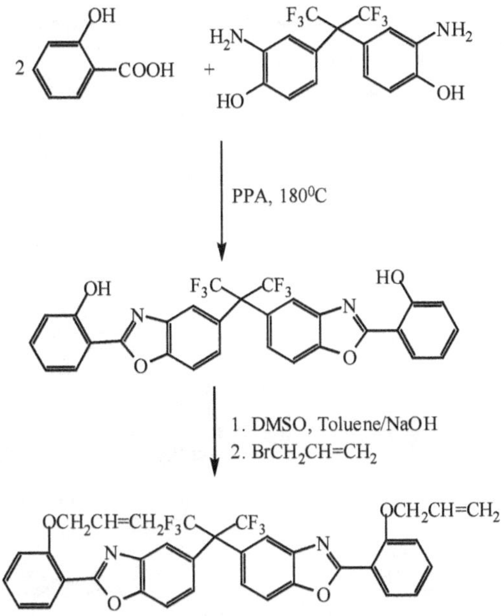

2 ⬡—COOH (OH) + (structure) PPA, 180°C

1. DMSO, Toluene/NaOH
2. BrCH$_2$CH=CH$_2$

(48)

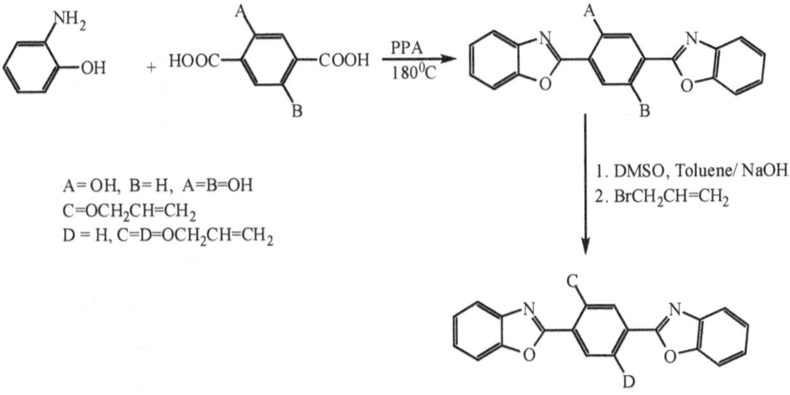

NH$_2$ / OH + HOOC—⬡—COOH (A, B) PPA 180°C

A= OH, B= H, A=B=OH
C=OCH$_2$CH=CH$_2$
D = H, C=D=OCH$_2$CH=CH$_2$

1. DMSO, Toluene/ NaOH
2. BrCH$_2$CH=CH$_2$

(49)

The oldest method in the synthesis of Benzoxazoles [23] (50) is heating or distilling 2-formamidophenols at elevated temperatures.

(50)

R = COOCH$_3$, COOH, Cl
R' = COOCH$_3$, NO$_2$

New bis (Benzoxazoles) (51, 52) have been synthesized in excellent yield from the corresponding bis (*o*-aminophenol) by refluxing with tri-ethyl orthoformate[24].

(51)

(52)

A novel series of 2-Aryldienylbenzoxazoles were prepared by Takatoshi Kosaka *et al.*[25]. The compounds (53) were prepared from *o*-aminophenols which were chloroacetylated, cyclized by ethyl polyphosphate and subjected to the Arbuzov reaction to give phosphonates. The condensation of aldehydes with phosphates by Hormer-Wadsworth-Emmons reaction and compounds were deprotected under acid conditions.

a= ClCH$_2$COCl, NaHCO$_3$, Acetone, rt   f= NIBAH, THF, -78°c

b= Ethyl polyphosphate, ClCH$_2$CH$_2$Cl, reflux   g = MnO$_2$,CH$_2$Cl$_2$, rt

c= (EtO)$_3$P, 150°C   h= NaH, THF -10°C

d= EtOH, H$_2$SO$_4$, reflux   i= NaOH,nBu$_4$NBr,H$_2$O, CH$_2$Cl, rt

e= MOMCl, ipr$_2$Net,CH$_2$Cl$_2$, rt   j=aq. 4M HCl, THF, rt

The one-pot thermal reaction of 1,3-bis (o-acylaminophenyloxy)-2-methylene propane derivatives gave the bis(benzoxazole) derivatives (54) via tandem Claisen rearrangement, in good yields [26].

2-amino-4-methyl phenols are treated with the corresponding acid chlorides in the presence of pyridine in dry DMF at 0-20°C for 12 hrs followed by etherification with 3-chloro-2-chloromethyl -1-propane in the presence of a base at 70°C for 12hrs to give bis (o-acylaminophenyloxy)-2-methylene propanes (54). The thermal bulk reaction of (54) gives the corresponding bis(Benzoxazoles) [27] (55).

Some new spiro-(pyran-4,2'benzoxazole) and derivatives were also synthesed [28].

Solid phase synthesis of benzoxazoles via Mitsunobu reaction was reported by Fengjiang Wang and James R. Hauske [29]. 2-Aminophenol attached to a solid support can be converted to the corresponding benzoxazole (56) by treatment with triphenylphosphine and diethylazodicarboxylate in THF at room temperature, in high yield and purity.

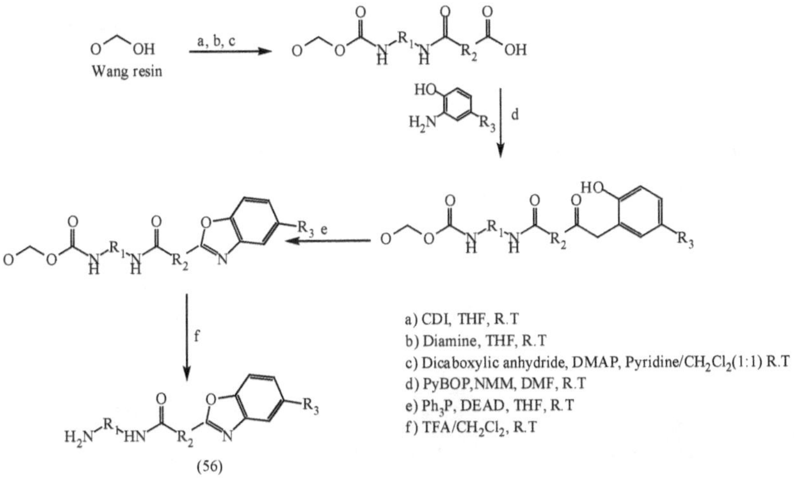

Wang resin
a, b, c
d
e
f
a) CDI, THF, R.T
b) Diamine, THF, R.T
c) Dicaboxylic anhydride, DMAP, Pyridine/CH2Cl2(1:1) R.T
d) PyBOP,NMM, DMF, R.T
e) Ph3P, DEAD, THF, R.T
f) TFA/CH2Cl2, R.T

(56)

The Aza-Witting reaction of the triphenyl phosphoranylideneamino-1,4-benzoquinone with aryl isocyanates and aryl chlorides allows the preparation of benzoxazole derivatives (57). The same reaction using aminophosphoranoquinone provides substituted benzoxazoles (58).

ArNCO
H2O

(57)

(58)

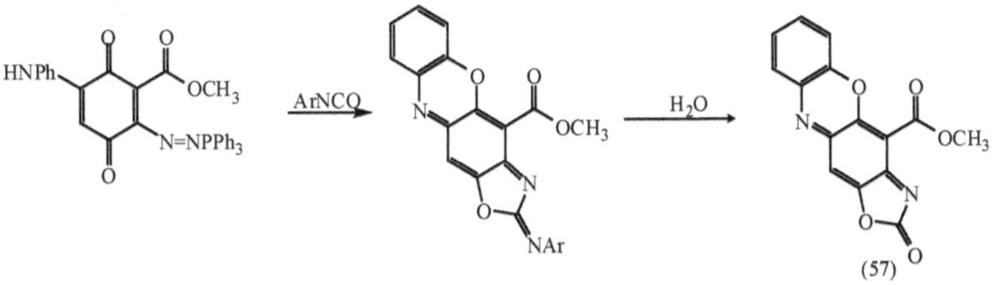

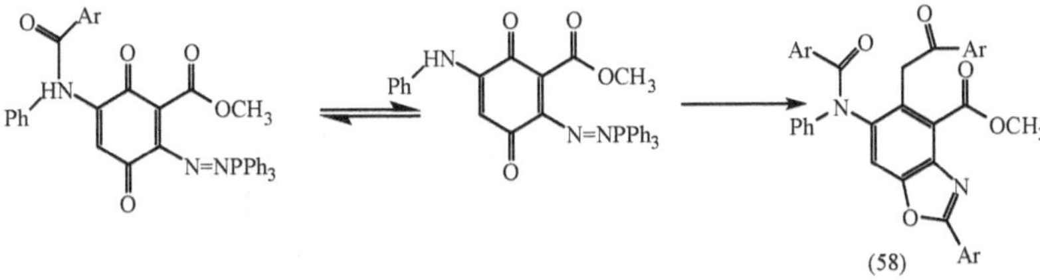

Ruthenium complex –Catalyzed facile of 2-substituted Benzoxazoles (59) reported by Kondo et al.[30]

RuCl$_2$ (PPh$_3$) $_3$ shows high catalytic activity for reaction of o-aminophenol with ROH (R = Bu, aralkyl) to give the corresponding benzoxazoles.

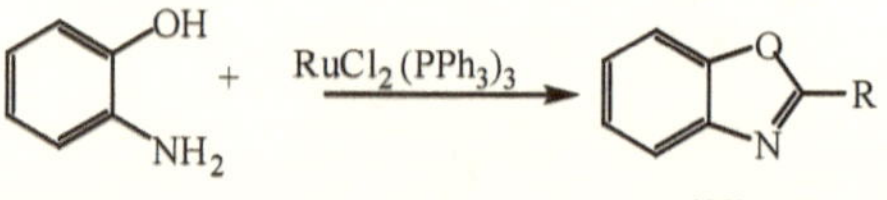

(59)

R = Bu, Aralkyl

Keten S,S-acetal reacts with 2-amino phenol to afford 2-(1-acetyl-2-oxopropylidene)benzoxazole (60), which was allowed to react with a variety of active methylens having an α-cyano of α –keto group to give spiro (pyran-4,2'-benzoxazole) derivatives.

Keten S,S-acetal obtained by reaction of acetyl acetone, CS$_2$ and two moles of methyl iodide in a one pot reaction using PTC [K$_2$CO$_3$/benzene/ tetrabutylammonium bromide (TBAB)].

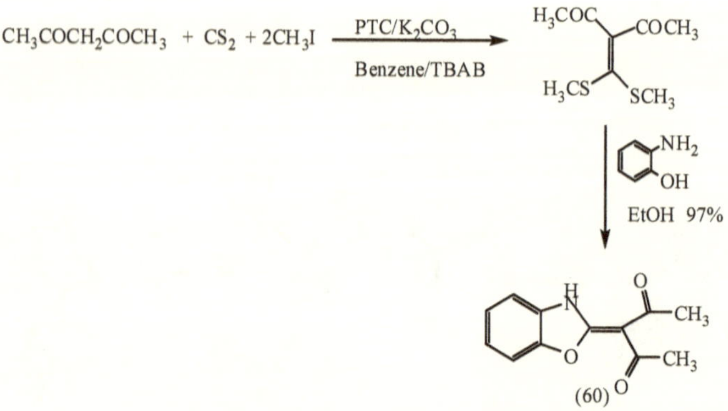

(60)

The simple and convenient synthesis of 5-substituted Benzoxazoles has been reported by Kenneth R.Kunz,[31].

5- Substituted Benzoxazoles (61) are prepared from 4-substituted -2-aminophenols by treatment with trimethyl orthoformate and concentrated aqueous hydrochloric acid.

(61)

R= Acetamido, Benzoyl, Bromo, Chloro, Cyano, Iodo, Methoxy, Methyl, Nitro, Propionyl.

Benzoxazole (1) is also obtained from the dry distillation of formamide and 2-aminophenol [32].

(1)

Phosphoryl methyl benzoxazoles (62) were prepared in three steps from o-aminophenols by 1) chloroacetylation with chloroacetyl chloride in presence of $NaHCO_3$. 2) Oxazole formation by treatment with ethyl polyphosphate and   3) Arbuzov reaction with triethyl phosphite [33]

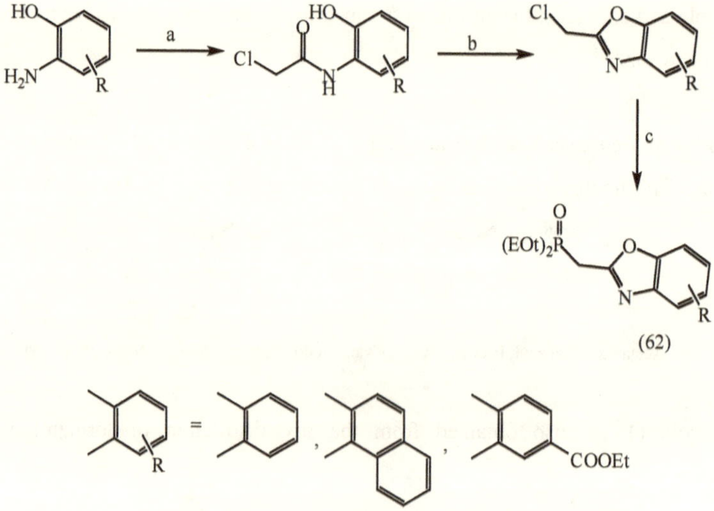

(62)

a = ClCH$_2$, COCl, NaHCO$_3$, acetone, r.t
b= Ethyl phosphate, ClCH$_2$CH$_2$Cl, reflux
c= (EtO)$_3$, 150°C

A one-pot synthesis of benzoxazoles carried out by chromium-manganese redox coupled reactions reported by Anitha Hari et al.[34]

The reaction in which a chromium-manganese redox couple is employed both to catalytically reduce an o-hydroxy nitroarene and to oxidatively cyclize a subsequently formed imine (63).

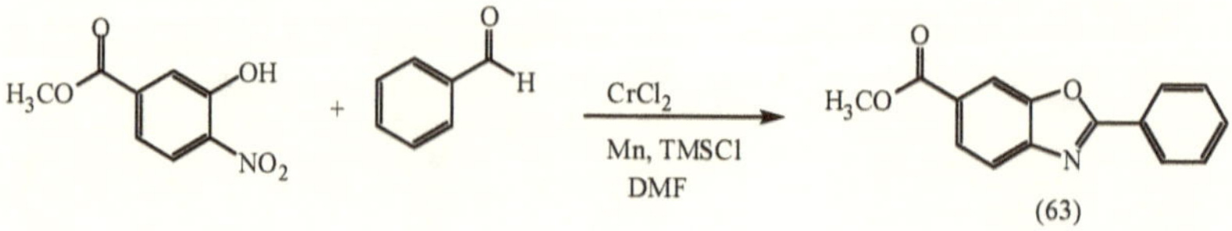

(63)

2-Hetero aromatic substituted isothiocyanatobenzoxazoles were synthesized by Haugwitz et al, [35]

The intermediates 5 and 6-nitrobenzoxazoles were synthesized employing the four parallel routes.

a) PPA-catalysed ring closure of o-aminophenols with the appropriate carboxylic acids followed by nitration of benzoxazoles.

b) Acylation of nitroaminophenols with carboxylic acid chlorides and subsequent thermally induced cyclodehydration of the amides.

c) Oxidation cyclization of Schiff bases using lead(IV) acetate and

d) Reaction of imino ethers derived from 2- cyanopyrazine or cyanonitropyridine with o-aminophenols and nitration of corresponding benzoxazoles.

Followed by thiocarbonylation of resulting amines using thiophosgene, completed the synthesis of 5 and/ or 6-isothiocyanatobenzoxazoles (64).

(64)

Salome Rodreguez Morgade et al. [36] reported the synthesis of benzoxazole derivatives.

Facile synthesis of 2-Substituted Benzoxazoles via ketens reported by Olagbemiro et al.[37]

The generation of diphenyl-, phenyl-, phenoxy- and chloroketens by treatment of corresponding acid chlorides with triethylamine in the presence of 2-aminophenol resulted in good yields of 2-substituted Benzoxazoles (65).

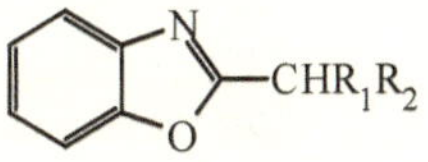

(65)

R_1, R_2 = Ph, Ph; H; PhO, H; Cl,H

Qian *et al.* [38] reported, yellow HgO as an efficient cyclodesulfurizing agent in the synthesis of 2-(substituted amino) Benzoxazoles from N-(2-hydroxy phenyl)-N'-phenylthioureas .

2-(Substituted amino) Benzoxazoles (66) (R_1, R_2 = H, F; R_3, R_4 = H, Cl, F; R_5= H, F, OH) were prepared in good yields by cyclodesulfurization of N-(2-hydroxyphenyl)-N'-phenyl thioureas with yellow HgO.

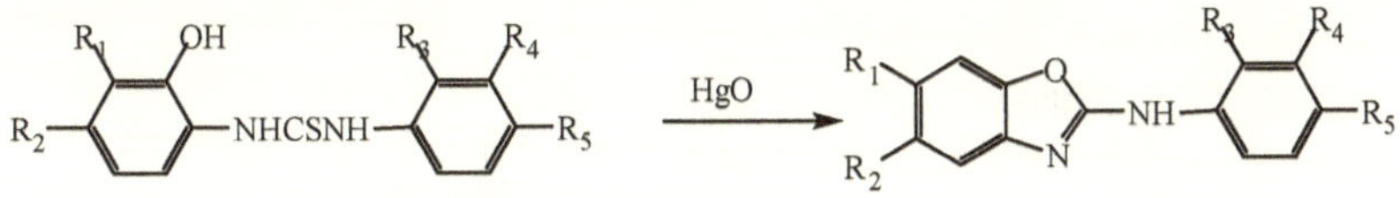

(66)

Aboulwafa *et al,* [39] reported the synthesis of several 2-ethoxycarbonyl Benzoxazoles (67). They were synthesized in high yields by cyclodesulfurization of the corresponding thioureas and thiosemicarbazide derivatives with dicyclohexylcarbodiimide (DCC).

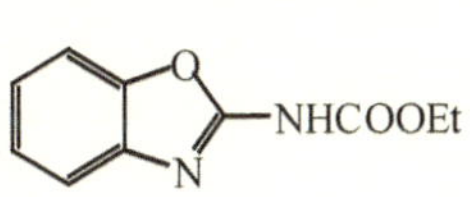

(67)

27

Summary of synthetic methods for the preparation of Benzoxazoles

Benzoxazoles are readily prepared from *o*-aminophenols and carboxylic acid derivatives as summarized below.

Heating is often required to cyclize an intermediate and yields range from moderate to excellent.

Another route of synthesis of 2-arylbenzoxazole utilizing the photochemical reaction or oxidative ring closure of imines.

o-nitroaroyloxybenzenes upon reduction with Sn-HCl yielding Benzoxazoles.

Benzoxazoles are also prepared by reacting *o*-hydroxyphenyl ketoximes with PCl_5 or P_2O_5.

A new synthetic route for preparation of Benzoxazoles, involves the reaction of *o*-amino phenols with allenic and acetylenic nitriles, leading to a range of Benzoxazoles in high yields.

Preparation of 2-oxo-, thio-, aminobenzoxazoles include the thermolysis of N-(o-hydroxyphenyl) urethanes, the reaction of o-aminophenol with CS_2, and the mercuric oxide oxidation of N-(o-hydroxy phenol)thiourea.

In a recent method treatment of N, O- diacylated-2-aminophenols with p-toluene sulfonic acid under reflux in xylene or toluene gives Benzoxazoles, in good yields.

Benzoxazoles are prepared by Palladium-catalyzed condensation of aryl halide with o-aminophenols followed by dehydrative cyclization.

The newer method for the preparation of Benzoxazoles makes use of the cycloaddition of o-hydroxyanilides by heating with aq.HCl at 150°C.

II. Biological Activities of Benzoxazoles

Targets containing the Benzoxazole moiety, either isolated from plants or accessed by total synthesis have remarkable biological activities [40]. For example Gram-positive antibacterials [41] polycyclic antibiotics [42], antiparasitic [43], anti-inflammatories [44], elastase inhibitors [45] and H_2-antagonists [46] containing the benzoxazole fragment.

Benzoxazoles have a number of optical applications such as photoluminescents [47], whitening agents [48] and dye laser [49].

The novel antibacterial agent containing the benzoxazole system is 'boxazomycin B' [50]. Benzoxazole ring containing antibiotic, calcymicin [51] (68) and the anti-inflammatory agent, benzoxaprofen [52] are also obtained by synthetic methods.

(68)

Zoxazolamine [53] (69), an α-amino-5-chlorobenzoxazole is reported to possess muscle relaxant, sedative activity and uricosuric effect.

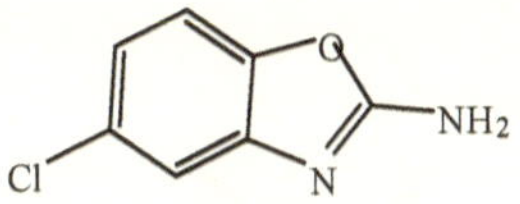

(69)

Chloroxazone (70) a chlorobenzoxazolidinone has skeletal muscle relaxant property and used for reduction of painful muscle spasms in medical and orthopedic disorders.

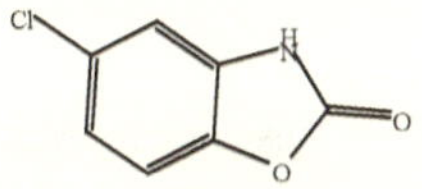

(70)

These examples highlight the level of interest in new synthetic approaches to benzoxazole derivatives.

These classes of compounds are considered to be important in view of their varied pharmacological properties. The following literature survey throws much light on the pharmacological significance of benzoxazole derivatives. For the sake of convenience, it is presented on the basis of the type of biological activity.

1. BENOXAZOLES AS ANTICANCER AGENTS

Benzoxazole (3,2-a)quinolinium salt (71) was synthesized and screened for biological activity was reported by Osvaldocox et al[54].

(71)

The compound showed DNA denaturation.

Hydrazone derivatives (72) are reported as anticancer and antimicrobial agents [55]

(72)

Some polycyclic fused benzoxazole derivatives (73) have been reported as potent anticancer agents reported by C.C.Cheng et al. [56]

(73)

In the benzoxazole series the most active compounds were found to be analogues with nitrile substituents on the phenyl ring (74). Methylation of the benzoxazole nitrogen leads to a decrease of activity for the 4-CN derivatives (75) reported by Eckhard *et al.* [57]

(74)   (75)

The synthesis of various derivatives of benzoxazole (76) displayed heparanase inhibition and angiogenesis reported by Martin *et al.* [58].

(76)

Various series of 2-substituted benzoxazole derivatives were synthesized [59] and evaluated for *in vitro* biological activity, some of these tested compounds (77 and 78) exhibited broad spectrum anti-tumor activity.

(77)   (78)

Four classes of UK-1 analogues were synthesized [60] and their cytotoxicity testing against human A-549, FTC- 905, RD, MES-SA, and HeLa carcinoma cell lines was determined, compound (79) is more potent than UK-1 against A-549 and HeLa cell lines.

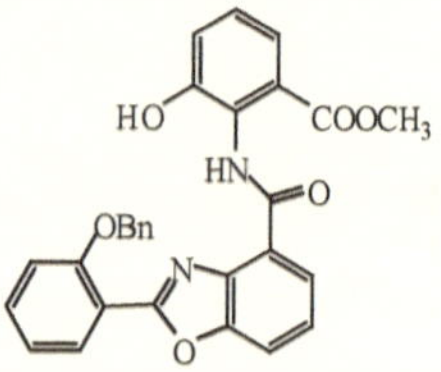

(79)

2. BENZOXAZOLES AS ANTIMICROBIAL AGENTS

Compounds (80, 81 and 82) were four times as active as streptomycin and about half that of ampicillin against the Gram-positive *Bacillus subtilis* (MIC < 12.5 μg ml^{-1}), while the activity of compounds (83and 84) was two times as active as of streptomycin and one fifth that of ampicillin, reported by Samia *et al.* [61].

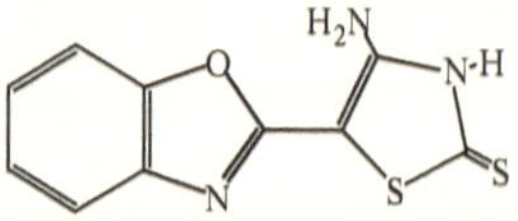

(80)

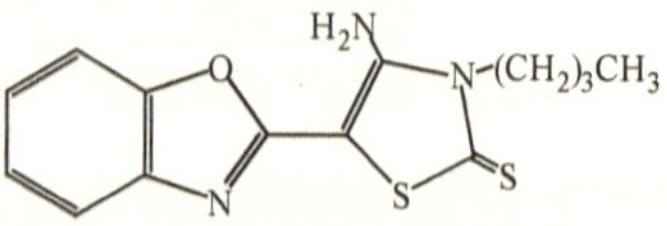

(81)

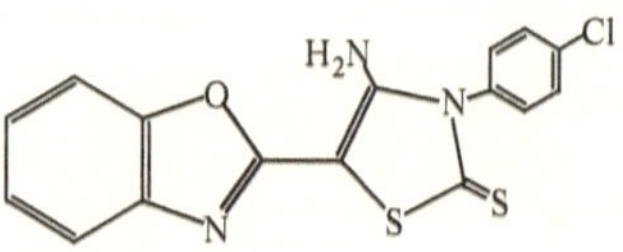

(82)

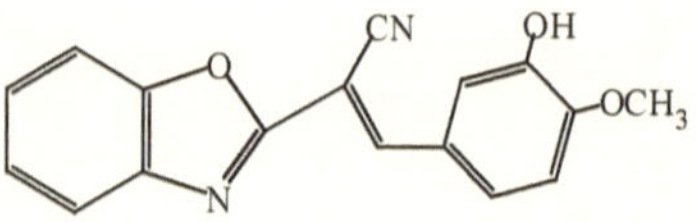

(83)

(84)

Some 2-[(α-methylbenzylidene)hydrazine]benzoxazole derivatives (85) were synthesized and their antimicrobial activities were investigated by the micro dilution susceptibility test in Mueller-Hinton broth and sabouraud liquid medium employing the test organisms; *Staphylococcus aureus* and *Enterococcus faecalis* as Gram Positive bacteria and *Candida albicans, Candida stellatoidea, Candida parapsilosis* and *Candida pseudotropicalis*, as yeasts. Among the compounds tested 2-[(α-methyl-4-chlorbenzylidene)-hydrazine] benzoxazole and 2-[(α-4-nitrobenzylidene)-hydrazine] benzoxazole showed the most favorable activity [62].

(85)

Some 2-(4-aryl-2-thiazolyl)-1,3-benzoxazoles (86) were synthesized and tested for their antifungal and antibacterial activities. All the compounds showed moderate activity against *A.niger, A.flavus, E.coli* and *S.aureus*[63].

(86)

35

Several 2-[[(benzoxazole-2-yl)sulfanyl]acetylamino]thiazole derivatives (87) were synthesized their antimicrobial activities against *Micrococcus luteus* (NRLL B-4375), *Bacillus cereus* (NRRL B-3711), *Proteus vulgaris* (NRRL B-123), *Salmonella typhimurium* (NRRL B-4420), *Staphylococcus aureus* (NRRL B-767), *Escherichia coli* (NRRL B-3704), *Candida albicans and Candida globrata* were investigated. The antibacterial assessment revealed that the compound possesses significant activity [64].

(87)

The synthesis of several new benzoxazole derivatives (88) showed the most prominent activity against various Gram positive and negative bacteria [65].

(88)

Seven 5-chloro-2(3H)-benzoxazolinone-3-acetyl-2-(p-substituted benzal-hydrazone derivatives (89) and four 5-chloro-2(3H)-benzoxazolinone-3-acetyl-2-(p-substituted acetophenone) hydrazone derivatives (90) were synthesized [66]. Their microbial activity done against two Gram positive bacteria (*Staphylococcus aureus, Bacillus subtilis*), two Gram negative bacteria (*Pseudomonas aeruginosa, Escherichia coli*) and two yeast like fungi (*Candida albicans, Candida parapsilosis*).

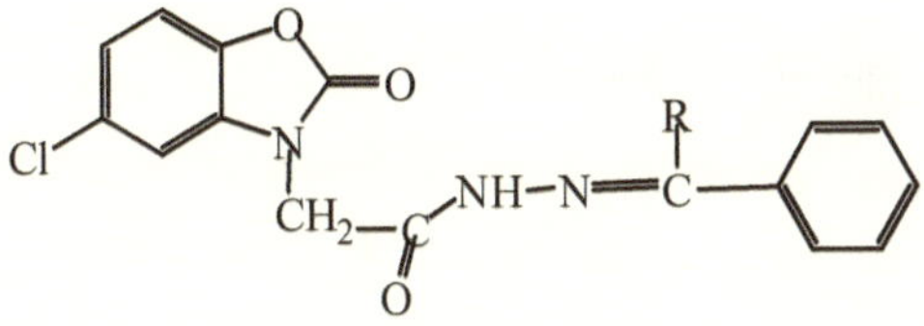

(89)

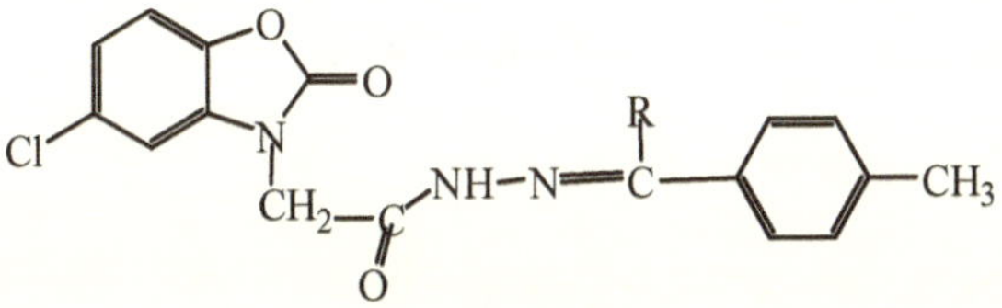

(90)

3. BENZOXAZOLES AS ANTI-HISTAMINIC AGENTS

A series of imidazo[1,2-a]pyridinylalkylbenzoxazole derivatives (91) were synthesized and tested for histamine H_2-receptor antagonistic, gastric anti-secretory and anti-ulcer activities. Some of 2-amino-6-[2-(imidazo[1,2-a]pyridine-2-yl)ethyl]benzoxazole and 2-acetamido-6-[2-(7-methylimidazol[1,2-a]pyridine-2-ylethyl]benzoxazole showed potent anti-secretory and cytoprotective activity [67].

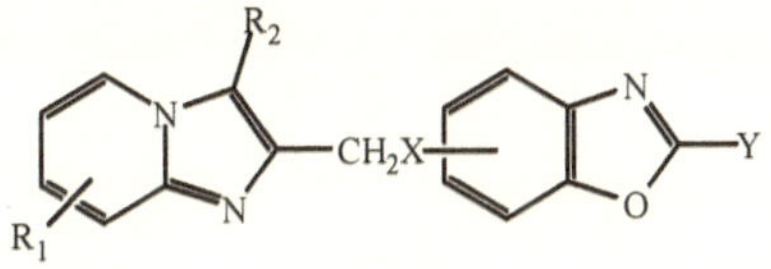

(91)

5-nitro-2-(p-substituted benzyl)benzoxazole derivatives (92) showed a good antihistaminic activity (92) reported by Nouanepalan et al. [68].

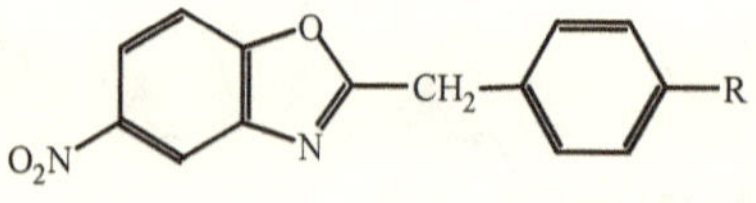

(92)

The preparation of Benzoxazole derivatives of the type (93) from ethyl 6-hydroxybenzoxazole-2-carboxylate and 2-(chloromethyl)quinoline, these compounds were found to be useful in treating allergic disorders reported by Jetsuya Makino *et al.*[69].

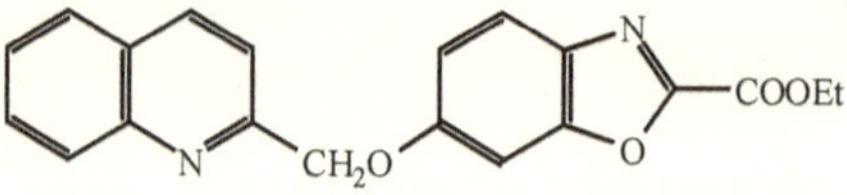

(93)

4. BENZOXAZOLES AS ANTHELMINTIC AGENTS

The synthesis of Isothiocyanato-2-pyridinyl benzoxazoles (94) [70] showed good parasitic activity.

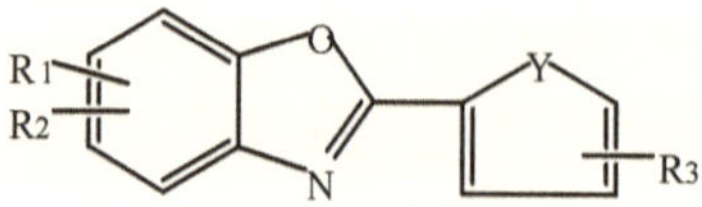

(94)

5. BENZOXAZOLES AS HYPOGLYCEMIC AGENTS

A new series of benzoxazol-2,4-thiozolidinones (96) was synthesized and evaluated for hypoglycemic activity in obese and diabetic yellow KK mice. 2-arylmethyl- and (2-

heteroarylmethyl)benzoxazole derivatives showed more potent activity than known 2,4-thiozolidinone derivatives such as ciglitazone, troglitazone and plioglitazone.

Only the 2-benzylbenzoxazole derivative showed excellent hypoglycemic activity, being more potent than all of the reference compounds. Decrease or increase in the length, or modification of the chain by cyclo alkyl or phenyl, branching and introduction of the oxo group, a double bond or triple bond all gave poorly active or inactive compounds. These finding led to synthesize a series of analogues of 2-benzylbenzoxazole derivatives. Substitution on the benzene ring in the lipophilic side chain also had a striking effect, and the results showed that the various substituents at the 4-position of the benzene ring such as fluoro, chloro, methoxy, ethoxy, phenyl, nitro and trifluoromethyl improved the activity.

In particular chloro, nitro and trifluoromethyl improved the activity 10 times more than 2-benzylbenzoxazole derivative. Although substituents at the 2- or 3 positions were less active than those at the 4-position, the 3,4-dichloro analogue showed very potent activity reported by Kenji Arakawa et al.[71].

(96)

6. BENZOXAZOLES AS CHYMASE INHIBITORS

The non peptide chymase inhibitors were designed on the structure of a peptidic compound (97) and demonstrate that the combination of a pyrimidinone skeleton as a P3-

P2 scaffold and heterocycles as P1 carbonyl activating group can function as nonpeptidic chymase inhibitor.

In particular, introduction of heterobicycles such as benzoxazoles resulted in more potent chymase-inhibitory activity. Substitution by an electron-withdrawing fluorine atom or an electron-releasing methoxy group at the 5-position of the benzoxazole ring resulted in a 4-fold more potent activity towards chymotrypsin and slightly more potent activity toward chymase than in the nonsubstituted analogue.

The introduction of a hydroxyl group led to a 10-fold decrease in both chymase and chymase trypsin-inhibitory activity, but neither an amino nor a phenyl group had any effect on chymase-inhibitory activity. The methoxycarbonyl analogue showed 5-fold greater inhibitory activity towards both chymase and chymotrypsin. The replacement with a carboxy or carbamoyl group resulted in loss of inhibitory activity, shifting the methoxycarbonyl group to the 6-position of the benzoxazole ring did not improve inhibitory activity.

Substitution on the benzoxazole ring also had no great effect on selectivity for chymase over chymotrypsin. This suggested that the substituents screened do not interact specifically with a part of the S' subsite but has some hydrophobic interaction with the S' subsite of both enzymes [72].

(97)

7. BENZOXAZOLES AS 5-HT₃ AGONISTS

Several modified 2-piperazinyl benzoxazole derivatives (98), which exhibit an agonist effect on gastrointestinal motility, were synthesized and their effects on the contraction of guinea pig ileum were examined. The quaternary piperzinyl benzoxazole structure has a restricted confirmation and stereostructure compared to those of the other 5-HT₃ receptor agonists. Serotonine and Metachlorophenyl biguanide. The mutual positions of the aromatic ring, nitrogen atom and terminal amine are considered to form the pharmacophore of the 5-HT₃ receptor agonist in the gut. The resulting suggested that, in these 5-HT₃ receptor agonists, the substituents of the benzoxazole ring influence the b-j reflex-inducing activity in rats [73].

(98)

A series of benzoxazole with nitrogen-containing heterocyclic substituent (99) at the 2-position was prepared and evaluated for 5-HT₃ partial agonist activity on isolated guinea pig ileum. The nature of the substituent at the 5-position of the benzoxazole ring affected the potency for the 5-HT₃ receptor, and the 5-chloro derivatives showed increased potency and lowered intrinsic activity. 5-chloro-7-methyl-2-(4-methyl-1-homopiperazinyl)-benzoxazole exhibited a high binding affinity in the same range as that of the 5-HT₃ antagonist granisetron, and its intrinsic activity was 12% of that of 5-HT. It inhibited 5-HT-evoked diarrhea but did not prolong the transition time of glass beads in the normal distal colon even at a dose of 100 times the ED₅₀ for diarrhea inhibition in mice. Compounds of this type are expected to be effective for the treatment of irritable bowel syndrome without the side effect of constipation [74].

(99)

Satoshi Yoshida *et al.*, [75] synthesized 2-(1-piperazinyl) benzoxazole and 2-(1-homopiperazinyl) benzoxazole derivatives and investigated their activities as 5-HT$_3$ receptor partial agonists. Compound (100) exhibited favorable profiles both *in vitro* and *in vivo*. This compound possessed not only high affinity and appropriate intrinsic activity for the 5-HT$_3$ receptor, together with high metabolic stability in an *in vitro* study, but also exhibited antidiarrhetic activity without preventing normal bowel function *in vivo*. These results suggest that compounds of this type are promising candidates for the treatment of irritable bowel syndrome, having satisfactory antidiarrhetic activity without causing constipation.

(100)

Gavin W.*et al.* reported that 2-Aminated Benzoxazoles (101) shows a good Selective 5-HT$_3$ receptor antagonists[76].

(101)

8. BENZOXAZOLES AS HIV-1-RT-INHIBITORS

The synthesis of non nucleoside HIV-1reverse transcriptase inhibitors containing, a CF_2 group. The SRN_1 reactions of 2-(bromodifluoromethyl) benzoxazole with the anions derived from heterocyclic thiols and phenolic compounds reported by Burkholder C.R et al.,[77]

(102)

3- [(benzoxazol-2-yl)ethyl-5-ethyl-6-methylpyridinyl-2(1H)- one] (103) (9, L-696,229), which was a highly selective antagonist of the RT enzyme (ICrn = 23 nM) and which inhibited the spread of HIV-1 IIIB infection by >95% in MT4 human T-lymphoid cell culture (CICgb = 50-100 nM), was selected for clinical evaluation as an antiviral agent[78].

(103)

9. BENZOXAZOLES AS HERBICIDAL AGENTS

Synthesis of some benzoxazole derivatives of type (104) showed good herbicidal properties [79].

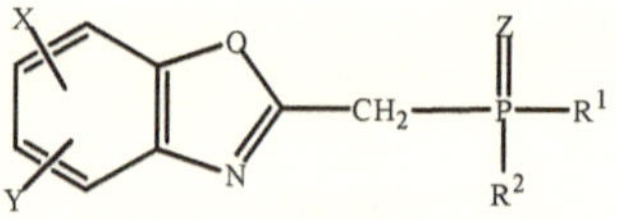

(104)

Synthesis, and QSAR's of some 5-substituted-2-(p-substituted benzyl) benzoxazoles (105) using Free-Wilson analysis possess antibacterial activity and good herbicidal properties [80].

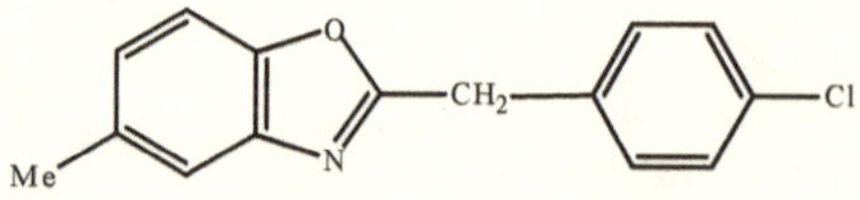

(105)

Antibacterial activity against *S. aureus* was determined using progressive double dilution technique. The compounds were found to be significantly active (MIC=6.25-50mg/ml). The multiple regression analysis also indicated that the 5-CH$_3$ and p-Cl groups are the most favourable substituents in compound (105).

Six benzoxazole derivatives (106) were prepared and assayed for their antifungal activity. There was a correlation between physicochemical parameters and antifungal activity *in vitro* [81].

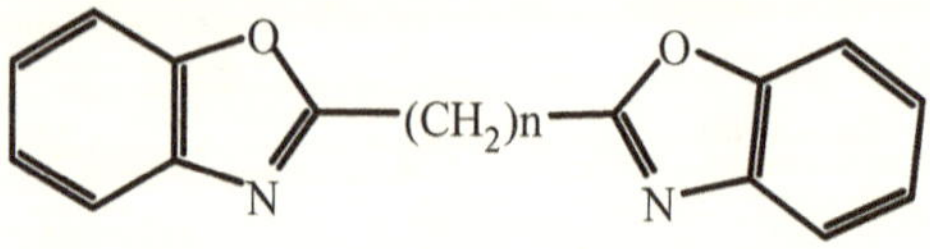

(106)

n=2 to 4

Synthesis of 5-amino-2-(*p*-substituted phenyl) benzoxazoles (107) were found to possess antimicrobial activity reported by Sener *et al* [82]. The MIC =6.25mg/ml for these compounds.

(107)

R=H, Br, F, Et, NMe$_2$, NO$_2$

The synthesis of 3-[4-(N-morpholino/piperino-methyl) phenyl-amino-methyl] benzoxazoles (108) showed good antibacterial activity [83].

(108)

R= H, Z=O
R=H, Z=CH$_2$
Both these compounds were found to exhibit antibacterial activity.

10. BENZOXAZOLES AS POTENTIAL AGONISTS OF LH HORMONES

The preparation of some 3-(2-benzoxazolyl)-D-alanine (109) by the condensation of *o*-aminophenol and β-carboxy function of α-benzyl-N-(benzyloxy carbonyl)-D-aspartate reported by John .J Nestor *et al.*[84] Incorporation of these amino acids into the 6th position of lutenizing hormone – releasing hormone (LH-RH) led to a series of potent

45

agonist analogues (upto 100 times LH-RH potency), active in doses ranging from 0.1 to 0.5µg by twice daily injection in rat oesterus cyclicity suppression assay designed to show the paradoxical anti fertility effect of these compounds.

(109)

11. BENZOXAZOLES AS URICOSURIC AND DIURETICS

A series of 7,8-dihydrofuro[2,3-g]benzoxazole-7-carboxylic acids (110) were synthesized and evaluated for uricosuric and diuretic activities in rats. They exhibited potent anti-diuretic activities with little effect on urate excretion reported by Haruhiko Sato et al.[85]

(110)

12. BENZOXAZOLES AS STEROID SULFATASE INHIBITORS

2-substituted Benzoxazoles which carry a sulfamic acid ester group board via oxygen to the p^H part of the ring structure more active (111)[86] .

(111)

They are indicated for use as steroid sulfatase inhibitors in the prevention, treatment of illness, responsive to alopecia, hisrutism, estrogen and androgen dependent cancer, inflammatory or autoimmune diseases, skin disorders or decreased congnitive function. The compound (112) with R_3 = adamant-2-ylidinemethyl was the most preferred agent of the invention.

(112)

14. BENZOXAZOLES AS INHIBITORS OF IMMUNO COMPLEX INDUCED INFLAMMATION.

The synthesis of 3-[1-(2-benzoxazolylhydrazine)-propionitrile derivatives (113)[87] were evaluated in the dermal and pleural reverse passive Arthus reaction, in the rat. In the pleural test, these compounds were effective in reducing exudates volume and accumulation of white blood cells.

(113)

47

14. BENZOXAZOLES HAVING MISCELLANEOUS ACTIVITIES

The benzoxazole system has been found to be an important building block for a number of light-stable fluorescent whitening agents, especially for their use on polyester fibers. The technically interesting 2-(4-phenylstiben-4'-yl) benzoxazole have been found to be obtained via the anil synthesis. Thus, reaction of 2-(p-tolyl)-5-tert-butylbenzoxazole with Schiff base from 4-formylbiphenyl and aniline in the presence of potassium hydroxide lead to 2-(4-phenylstilben-4'-yl)-5-tert-butyl benzoxazole (114). Similarly, with the corresponding anil from 4-formyl-p-terphenyl under the influence of potassium tert-butoxide, compound was obtained reported by Katritz et al., [88].

(114)

Twelve benzoxazole derivatives and a reference compound, 4(5)-phenylimidazole were tested against APDM and AH activities in hepatic microsomes from PB treated rats. Benzoxazole (1) was a weak inhibitor of APDM although it has previously been found to enhance the N-demethylation of p-chloro-N-methylaniline in rat liver microsomes. For the 2-alkylbenzoxazole series inhibitory potency towards APDM activity increased as the number of carbon atoms in the alkyl chain increased. The inhibitory potency of the parent compound, benzoxazole (compound 1, APDM 150=1X10-3M) was increased 26 times by 2-alkyl substitution. The most potent 2-alkylbenzoxazole (115) derivative tested against APDM activity was 2-n-heptylbenzoxazole reported by Peter J. et al.[89]

(115)

Peptidomimetic derivatives (116) featuring a P_1-arginonylheterocycle were designed. The preparation of two key building blocks containing benzoxazole ring and their incorporation into thrombin and factor X_a specific sequences was described. The serine protease inhibitory activity of these targets was evaluated. Models of the compounds complexed with thrombin have been constructed from a crystal structure of thrombin.

Placement of the compound in the active site of thrombin in fashion analogues to the peptidyl-α-ketobenzoxazoles in elastase produced favorable thrombin interactions of the benzoxazole NH and O with His 57 and Gly 193, respectively.Which would also be possible in trypsin and factor X_a placement of the compound in the active site of thrombin in the same fashion presents the favorable His 57 interaction. These compounds interesting levels of biological activity with varying selectivities towards related serine proteases [90].

(116)

Preparation and formation of 2-[5-9adamantyloxymethyl)-2-cyclohexyl-1H-imidzol-4-yl] benzoxazoles (117) as gastrin and cholecystokinin receptor ligands for treatment of G.I disorders [91]. The compound showed gastrin (CCK_2) antagonist activity. The compounds comprising a proton group inhibitor for treatment of gastro intenstinal disorders are described. These compounds reduced hyperplasia associated with administration of a proton pump inhibitor alone.

(117)

Synthesis of 2-arylbenzoxazoles (118) and investigation of their mutagenicity reported by Ucucu et al.,[92]. Eight of 2-arylbenzoxazoles were synthesized, mutagenicity tests (AMES test) of these carried out, no significant mutagenicity was observed.

(118)

2-(1-Methyl-1,2,3,6-tetrahydropyridin-4-yl) Benzoxazoles (121) were synthesized, and found to exhibit moderate binding affinities to the cloned [93]

(121)

Benzoxazolone (2-hydroxybenzoxazole) induces a sleeping state when administered hypodermically to mice but exerted no such action when given, benzoxazolone is relatively non-toxic since it has been isolated as the hydrolytic product of an unknown substance obtained from the urine after feeding formanilide or acetanilide to dogs, 2-ethylbenzoxazole (122) also possesses mild hypnotic activity, therefore, synthesized eight higher homologs of 2-ethylbenzoxazole as well as four aryl substituted derivatives and tested them for anticonvulsant activity[94].

(122)

R= Ethyl, Hydroxy, Heptyl, Benzyl, Phenyl, Pentadecyl, Hepta decyl etc.

Synthesis of a series of 2-benzylsulfanyl derivatives of benzoxazole (123) and their anti tuberculosis activities *in vitro* against sensitive and drug resistant strains is presented [95].

(123)

References:

1. Barton and Ollis, *Comprehensive Organic Chemistry*, Pergamon Press, **4** (1972) 962.

2. Megumi Yamada, Yasuo sato, Kazuko Kobayasui, Pukio Konno, Tomoko Soneda and Takashi Watanabe, *Chem. Pharm. Bull.*, **46(3)** (1998) 445.

3. Mathews, J. Craig, Clegg William, Elsegod, Mark R.J. Leeie, Troy A., Thorp Derek, Thorantan Peter, Lockhert, C. Joyee, *J. Chem . Soc., Dalton Trans,* **8** (1996) 1531.

4. T.Arun kumar and Jaya Prasad Rao, *Indian J. Heter. Chem.*, **11(1)** (2001), 9.

5. T.Sambaiah and K. Kondal Reddy, *Synthesis, Indian J. Heter. Chem.*, (1990) 422.

6. Jozsef Kover, Tibor Timar, Tozsef Tompa, *Synthesis Nov* (1994) 1124.

7. Katritzky and Rees, *Comprehensive Heterocyclic Chemistry,* **6**(1994)**,** 216.

8.   C.D.Hein, R.J. Alheim and J.J. Leavitt, *J. Am. Chem. Soc.,* **79** (1957), 427.

9.   Hofmann A.W and Miller W.V, *Ber.,* **14** (1881) 567.

10.   Guillaume Poissonnet, *I Synth. Commun.,* **27(22)** (1997) 3829.

11.   B.M. Bawal, S.P. Nayabhate, A.P. Likhile and A.R.A.S. Deshmukh,   *Synth. Commun.,* **25(12)** (1995) 3315.

12.   Rostamizadeh, Shohnez, Derafshian, Esmaiel, *J. Chem. Res. Synapses,* **6** (2001) 227.

13.   C.Grundmann and A. Kreutzberger, *J. Am. Chem. Soc.,* **77** (1955) 6559.

14.   G.L. Jenkins, A.M. Knevel and C.S. Davis, *J. Org. chem..,* **26** (1961) 274.

15.   M.Roussos and J.Lecomte, *German Patent* **1(124),**(1962) 499,*Chem. Abstr.,* **57** (1962) 9858.

16.   Y.Ito, Y. Inubushi, M.Zenbayashi, S. Tomita and T. Seegnasa, *J. Am. Chem. Soc.,***95** (1973) 4447.

17.   Y. Ito, I. Ito, T. Hirao and T. Saegusa, *Synth. Commun.,* **4** (1974) 97.

18.   J.P.Ferris and F.R. Antonucci, *Chem. Commun.,***126** (1972).

19.   R.H.Khan and R.C. Rastogi, *Indian J. Chem.,* **28B** (1989) 529.

20.   Yasuo Sato, Megumi Yamada, Sathoshi Yoshida, Tomoko Soneda, Midori Ishikawa, Tetsutaro Nizato, *J. Med. Chem.,* **41** (1998) 3015.

21.   Claudio Saitz, Herman Rodriquez, Amelia Marquez, Alvaro Canete, Carolina Jullian and Antonia Zanocco, *Synth. Commun,* **31(1)** (1998)**,** 135.

22.   Thuy D. Dang, Leslie S. Hudson, Willian A. Feld and Fred E. Arnold , *Polymer Preprints,* **41(1)** (2000), 103.

23.   Jois, H.R. Yajunarayana Gibson, W. Harry, *J. Heterocycl. Chem.,* **29(5)**, (1992) 1365.

24.   Jian- Guo Sheo Qi Zhong Hai Ping Liao, Chang Qing Liu and Jing-Ferg Zhou, *Org. prep proceed Int.,* **24(5)** (1992) 520.

25. Takatushi Kosaka, Keiko Ochiai, Setsuya Ohba, Toshio Wakabaysahi and Sei-itsu Murota, *Bio-org and Med. Chem. Letter,* **5(1),** (1995) 35.

26. Emiko Koyama, Garig Yang and Kazuhisa Hirdani, *Tetrahedron Letts,* **41** (2000) 8111.

27. A.Lodengberg, *Ber.,***10** (1877) 1123.E.Banberger, *Ber.,***36** (1903) 2042.

28. A.K. El-Shafei, A.M.M. El- Saghier, E.A. Ahmed, *Synthesis Fes,* (1994) 152.

29. Fengjiang Wang and James R. Hauske, *Tetrahedron Letters.,* **38(37)** ((1994)6529.

30. Kondo, Teruyuki, Yang, Sunborg, Huh, Keun Tae, Kobayashi, Masanobu, Kotachi, Shinji, Watanabe, Yoshihisa, *Chem. Lett.,* **7** (1991) 1275.

31. Kenneth R. Kunz, *Organic Prep & Proc. Int.,* **22(5)** (1990) 613.

32. S. Van Niementowski, *Ber.,* **30** (1897) 3062.

33. Takatoshi Kusaka and Toshio Wkabayashi, *Heterocycles,* **41(5)** (1995) 447.

34. Anitha hari, Charles karan, Warren C. Rodrignes and Benjamin L. Miller, *J. Org. Chem.,* **66** (2001) 991.

35. R.D. Haugwitz, R.G. Angel, G.A. Jacobs, B.V. Maurer, V.L. Narayanan, L.R. Cruthers and J. Szanto , *J. Med. Chem.,* **25** (1982) 969.

36. Salome Rodriguez-Morgode Purification Vazquez and Tomas Torres, Copy right (1996), 50040-4020 (96) 00290-6, 6781.

37. T.O. Olagbemiro, M.O. Agho, O.J. Abayeh, J.O. Amuptitan, *Recl. Trav. Chim. Pays-Bas.,* **115 (6)** (1996) 337.

38. Qian, Xuhong, Li, Zhibin, Sorg, Gonghua, Lizhorg, *J. Chem. Res. Synop.,* **4** (2001) 138.

39. Aboulwafa, M. Omaima, A. Omar, M.E. Mohsen, *Sulfur Lett.,* **14(4)** (1992) 181.

40. G.V. Boyd, In: *Comprehensive Heterocyclic Chemistry*, A.R. Katrizky, C.W. Rees eds. Pregamon Oxford, **64(B)** (1984) 178.

41. (a) T. Kusumi, T. Ooi, M.R. Walch, H. Kakisawa, *J. Am.Chem. Soc.,* **110** (1988) 2954.

(b) M.J. Suto and W.R. Turner, *Tetrahedron Lett.,* **36** (1995) 7213.

45.　(a) M.O. Chaney, P.V. Nemarco, N.D. Jones, J.R. Occolowitz, *J. Am.*

　　　Chem. Soc., **96** (1974) 1932.

　　　(b) L. David and A. Dergomard, *J. Antibiotics*, **35** (1982), 1409.

　　　(c) J.W. Wertly, J.W. Liu, J.F. Blount, L.H. Sello, N.Troupe and P.A.

　　　Miller, *J. Antibiotics,* **36** (1983) 1275.

46.　(a) R.P. Hangwitz, B.V. Maurer, G.A. Jacobs, V.L. Narayanan, L.R

　　　Cruthers and J.Szanto, *J. med. Chem.,***22**(1979) 1113.

　　　(b) R.P. Hangwitz, R.G. Angel, G.A. Jacobs, B.V. Maurer,

　　　V.L.Narayanan, R.L. Cruthers and J. Szanto, *J. med. Chem.,* **25** (1982)　969.

47.　(a)　D.W. Dunwell, P.Evans, T.A.Hicks, C.H. Cashin, A.Kitchen,

　　　*J.Med Chem.,***18**(1975) 53.

　　　(b) D.W. Dunwell, P.Evans, T.A.Hicks, *J.Med. Chem.,***18**(1975) 1158.

　　　(c) D. Evans, C.E. Smith, W.R.N. Williamson, *J.Med.　Chem.,***20**(1977)

　　　169.

　　　(d) D.W. Dunwell and P.Evans, *J, Med.Chem.,* **20** (1977) 797.

48.　(a) P.D. Edwards, E.F. Meyer, J. Vijayalaxmi, P.A. Tuthill, D.A.

　　　Andisik, B. Gomes, A. Strimplet, *J. Am. Chem. Sco.,* **114**(1992) 1854.

　　　(b) P.D. Edwards, J.R. Damewood, G.B. Steelman, C.Bryant, B. Gomes and J.Williams, *J. Med. Chem.,* **38** (1995) 87.

(c) P.D. Edwards, M.A. Zottola, M.Davis, J.Williams and P.A. Tuthil *J. Med. Chem.,***38**(1995) 3972.

49. (a) Y. Katsura, S. Nishino, Y. Inone, M.Tomoi and H.Takasugi,

Chem.Pharm.Bull., **40** (1992) 371.

(b) Y. Kastura,Y. Inoue, S. Nishno, M. Tomoi, H. Itoh and H. Tkasugi,

Chem. Pharm. Bull., **40** (1992) 1424.

50. U. Claussen and H. Harnisch, *Eur. Pat. Appli.,* **25** (1981) 136.

51. Stendby. S. *Surfactant Sci. Ser.,* **5** (1981) 729.

52. (a) B.M. Trost, I. Fleming. *'Comprehensive organic Chemistry'* ed.s C.H. heathcock, Ed. Pergamom Press, New York, **2** (1991)147.

(b) A. Reser, L.J. leyshon, D. Saunolers, M.V. Mijovic, A. Bright and *J. Bogie, J. Am. Chem. Soc.,* **94** (1972) 2414.

53. M.J. Suto and W.R. Turner, *Tetrahed. Lett.,* **36** (1995) 7213.

54. D.A Evans, C.E. Sacks, W.A. Kleschick, T.R. Taser, *J. Am. Chem. Soc.,***101**(1979) 6789.

55. D.W.Dunwell, D. Evans, T.A. Hicks, C.H. Cashin, and A. Kitchen. *J. Med. Chem.,* **18** (1975) 53.

56. Barton & Ollis, *Comprehensive Organic Chemistry,* **4** (1979) 962.

57. Osvaldo Cox, Henry Jackson, Vanessa A. Vargas, Adriana Baez, Julio I. Colon, Balacac C. Gonalez and Marino de Leon, *J. Med. Chem.,* **25**

(1982), 1378.

58. I.H. Hall, N.J. Peaty, J.R. Henry, J. Easmon, G. Heinisch and G.

Purshinger, *Arch. Pharm. (Weinheim)* **332** (1999) 4.

59. C.C. Cheng, D.E. Liu and T.C. Chou, *Heterocycles* **35(2)** (1993) 775,

Chem. Abstr. **120** (1994), 217507b.

60. Eckhard Bastona and Frederic R. Leroux ,*Recent Patents on Anti-Cancer*

Drug Discovery, **2**(2007)*,* 31-58.

61. Martin, C. S., Andrew, H. P. and Carter, S. D. I. *Current Bioactive*

Compounds, **1**(2005)*,* 1-24.

62. Mireya L. McKee, Sean M. Kerwin, *Bioorganic & Medicinal Chemistry*,

16 (4) (2008), (1775-1783).

63. Shu-Ting Huang, I-Jen Hsei, Chinpiao Chen, *Bioorganic & Medicinal*

Chemistry,   **14 (17)1**(2006), (6106-6119).

64. Samia M. Rida, Fawzia A. Ashour, Soad A.M. El-Hawas, Mona M.

ElSemary, Mona H. Badr and Manal A. Shalaby 120(2005), 72-75.

65. Seyam Ersan, Sultan Necok, Rukiye Berkem and Tuncel Ozden,

Arzneim- Forsch/Drug Re., **47(II) (8)** (1997)963.

66. R.H.Kahn and R.C. Rastogi, *Indian J.Chem,* **28B** (1989), 529.

67. Zafer Asim Kaplancikli, Gulhan Turan-Zitouni, Gilbert Revial, and

Klymet Guven *Arch Pharm Res.,* **27(11)** (2004) 1081-1085.

68. Tugba Ertan, Ilkay Yildiz, Betul Tekiner-Gulbas, Kayhan Bolelli, Ozlem

Temiz-Arpaci, Semiha Ozkan, Fatma Kaynak, Ismail Yalcin and Esin

Aki, *European J.of Med.Chem* **44(2)** (2009) 501-510.

69. Tijen Onkol Mehtap Gokce, AH Ulvi Tosun, Serpil Polat, Mehmet S.

Serin,*Turk J.Pharm.Sci.***5 (3)**(2008), 155-166.

70. Yousuke Katsura, Yoshikazu Inoue, Signetaka Nishino, Masaaki Tomoi,

Harunobu Itoh and Hisashi Takasugi, *Chem. Pharm. Bull.,* **4(6)** (1992)

1924.

71. Singur Noyanalapam and Esin Sener, *FABARD Farm Biimer Derg Turkish,*

11 (1986) 111; *Chem. Abstr.,* **106** (1987) 196302Z.

72. Tetsuya Makino, Tetsuya Kato, Takayuki Imaoka and Masayuki Kaneko,

Jpul Kokai, Tokyo JP 0446, 177 (92, 46, 177) (Cl. 07 D 413/12), 17 Feb.

1992; *Chem. Abstra.,* **117** (1992) 69856k.

73. R.D Haugwitz, R.G.Angel, G.A Jacobs, B.V. Maurer, V.L.Narayana,

L.R.Cruthers and J. Szanto *J.Med.Chem* **25** (1982), 969-974.

74. Kenji Arakawa, masanori Inamasu, mamoru Matsumoto, Kunihsto

P kumura, Konsuke Yasuda, Hindenori Akatsuka, Sobura Kawanani,

Akishige, Watanase, Koichi Homma, Yutaka Saiga, Masakatsu Jzei and

Ikuo Iijima, *Chem. Pharm.Bull.*, **45 (12)** (1997) 1984.

75.   Fumihiko Akahoshi, Atsuyuki Ashimori, hiroshi Sakashita, Takuya
      Yoshimura, Teruki Imada, masahide Nakasima, naoko Mitsutomi,
      Shiegeki Kuwahara, Tatsuyukioutsuka, Chikara Fukaya, Mizu Mixazaki
      and Norifumi Nakamura, *J. Med. Chem.*, **44** (2001) 1286.

76.   Megami Yamada, Yasuo Sato, Kazuko kobayashi, Fukio Konno, Tomoko
      Soneda and Takashi Watanasa, *Chem. Pharm. Bull.*, **46(3)** (1998) 445.

77.   Yasuo Sato, Megumi Yamada, Sathoshi Yoshide, Tomoko Soneda, Midori
      Ishikawa, Testutaro Nizato, Kokichi Suzuki and Fukio Konno, *J. Med.
      Chem.*, **41** (1998) 3015.

78.   Satoshi Yoshida, Sojiro Shiokawa, Ken-Ichi Kawano, Tomoko Ito,
      Hiroshi Murakami, Hisashi Suzuki, and Yasuo Sato *J. Med. Chem.*, *48*,
      *(*2005) 7075-7079.

79.   Gavin W. Stewart, Carl A. Baxter, Ed Cleator, and Faye J. Sheen *J. Org.
      Chem. Vol. **74**, No. **8**, (2009)* 3231.

80.   C.R. Brukholder, W.R.Dolbier, M. Medebielle, *J. Flourine Chem.* **102 (1-
      2)** (2000), 369.

81.   Jacob M. Hoffman, Anthony M. Smith, Clarence S. Rooney, Thorsten E.

Fisher, John S. Wai, Craig M. Thomas, Dona L. Bamberger, James L.

Barnes, Theresa M. Williams, *J. Med. Chem.,* **36 (8)** (1993) 953–966.

82. Peter Paul Wilhelm, Wilhelm, Sittenthales, Hans Ulrich Bernhard and

    Torsten Rehm, *Ger. Offen D.E.*3, 638685 (Cl. A. 01 N57/08), 1980,*Chem.*

    Abstr., **109** (1989) 110657.

83. Sener, Esin, Temiz, Oezlem, Oeren, Ilkay, Yalcin, Ismail, Akin, Ahmet,

    Ucartwerk, Nejat, *Ankara Univ. Eczacilik. Falc. Derg,* **24(1)** (1995) 10.

84. Cakir, Bilge, Ucucu, Umit, Buyukbingol, Erden, Abbasoglu and Ufuk,

    Gazi. Univ. Eczacilik Fak Derg., **6(1)** (1989) 15.

85. Sener, Esin, Yalcin, Iswail, Ozden, Seckin, Ozden, Tuncel, Akin, Ahmet,

    Yildiz, and Sulhiye, *Doga. Tip Eczacilik,* **11(3)** (1987) 391.

86. Rajendra S. Varma and Kaushal Verma, *Indian J. Chem.,* **25B** (1986) 877.

87. John J. Nestor, Jr., Bonnie L-Homer, Teresa L. Ho., Gordon H. Jones, Georgial' Me Rao and Brain H. Vickeny, *J. Med. Chem.* **27** (1984) 320.

88. Haruhiko Sato, Takoshi Dan, Etsuro Onuma, Haruko Tanaka, Bunya Aoki and Hiroshi Koga, *Chem. Pharm. Bu.,* **39 (7)**(1991) 1760.

89. Billich, Andreas, Schreiner, Erwin Paul, Wolf-Winiski, A.G. Barlara Noveris, GB 1999-27439 (1991), G.B. 2000-7511 (2001) *Patent CA Section*: **28** Section **1,2, 32**.

90. Forutna Haviv, James D. Ralajakezyk, Robert W. Denet Francis A. Kerdesky, Roland L. Walters Steven P. Schmidt, James H. Holms, Patrick R. Young and George W. Carter, *J. Med. Chem.,* **31** (1988) 1719.

91. Katritzky and Boulton, *Advances in Heterocyclic Chemistry,* **23** (1978) 202.

92.   Peter J. Little and Adrian J. Ryan, *Biochem. Pharma.*, **31(9)** (1982) 1795.

93.   Susan Y. Tamura, Brian M. Shamblin, Terenne K. Brunck and William C. Ripka, *Biorg. Med. Chem. Lett.*, **7(10)** (1997) 1359.

www.ingramcontent.com/pod-product-compliance
Lightning Source LLC
Chambersburg PA
CBHW021920170526
45157CB00005B/2110

9781329087859